Illustrated by Sally Kindberg

Photography by Peter Anderson, Paul Bricknell, Geoff Brightling, Jane Burton, Peter Chadwick, Andy Crawford, Geoff Dann, Mike Dunning, Neil Fletcher, Martin Foote, Steve Gorton, Frank Greenaway, Colin Keates, Dave King, Cyril Laubscher, Ray Moller, Tracy Morgan, Stephen Oliver, Susanna Price, Karl Shone, Steve Shott, Kim Taylor, Jerry Young

The publisher would like to thank the following for their kind permission to reproduce their photographs:
a = above, c = centre, b = below/bottom, l = left, r = right, t = top.

Night
Robert Harding Picture Library: Richard Pharaoh 6–7; **Oxford Scientific Films:** M.P.L. Fogden 9br; **Pictor International:** 3b, 5, 11tl; **The Stock Market**: Photo Agency UK 4–5; **Tony Stone Images:** James Balog 7c; Cosmo Condina 11br; Zigy Kaluzny 10br.

Gatefold
Oxford Scientific Films: Rob Nunnington tcl; **PowerStock/Zefa:** cr; **Tony Stone Images:** Silvestre Machado l.

Day
Ardea London Ltd: Ferrero-Labat 9cl; **Colorific!:** Michael Yamashita 5tl; **Pictor International:** 6b; **The Stock Market:** Photo Agency UK 11tl; **Tony Stone Images:** 4–5b; Gary Braasch 7tl; Mike McQueen 3c; Laurence Monnerat 11tl.

Jacket
Robert Harding Picture Library: Richard Pharaoh *moon*; **The Stock Market:** Photo Agency UK *house*; **Tony Stone Images:** *field.*

Contents

Night index

NIGHT

Discover the world in the darkness of night

Claire Llewellyn

DK

London • New York • Sydney • Moscow • Delhi

The evening is the start of a long, dark night.

Night hunter

At dusk, owls leave their daytime roosts and set out to hunt for food.

Soft wing feathers make no sound.

Twilight time

As night falls, the sky grows darker and the Moon comes out. The air begins to cool.

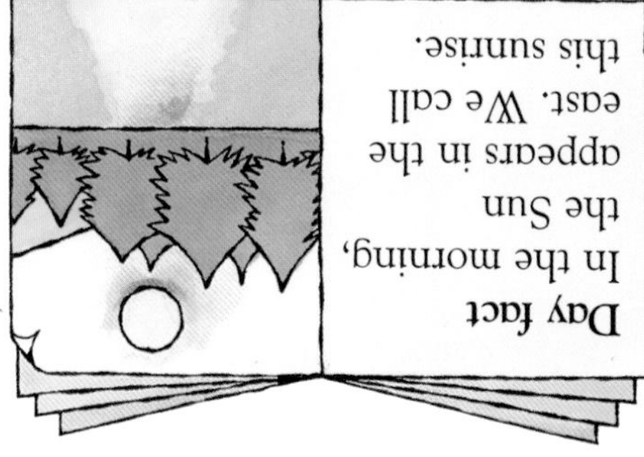

Good night!

By the evening, many animals are ready for a night's sleep.

City lights

Lights come on as the daylight fades. City skyscrapers shine brightly in the dark.

The Moon and twinkling stars light up the night sky.

Half moon

Crescent moon

Moon gazing

The Moon seems to change shape each night because different parts of it are lit up by the Sun.

Binoculars help you to see distant stars.

Comets

Sometimes, bright balls of ice, called comets, cut across the night sky.

Night bloomer

This cactus opens its white, scented flowers at night.

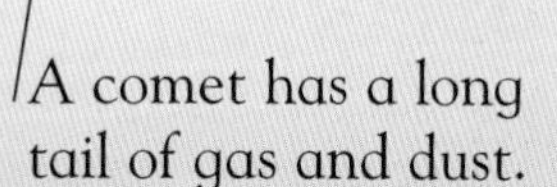

A comet has a long tail of gas and dust.

White petals stand out in the dark.

Night-time animals have sharp ears and huge eyes.

Hunting by hearing
Bats hunt by listening for echoes that bounce off flying insects.

Night feed
Snails come out when the air is cool and damp.

Big eyes

A bushbaby has eyes like saucers. It can see well at night.

Night fact
A glow-worm flashes a light at night, which helps it attract a mate.

Day fact
A bird of paradise has bright feathers to help it attract a mate.

Shiny eyes

A raccoon's eyes have a shiny layer. Any light makes them glow in the dark.

Weary people sleep and dream the night away.

Story time
Reading a story is a good way to relax before bed.

A good night's sleep
Sleep rests every part of the body. It gives us energy and helps us to think.

Night fact
A fireworks display is great after dark. It lights up the starry sky.

Day fact
A street parade is fun to watch. It has colourful costumes, flags, and floats.

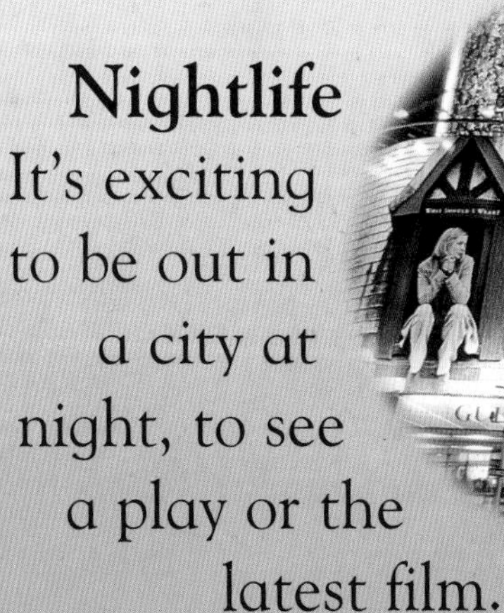

People lie down to sleep.

Night workers

Hospitals are always busy. Doctors and nurses work through the night.

Nightlife

It's exciting to be out in a city at night, to see a play or the latest film.

Sunset
In places near the equator, the Sun sets at the same time every day.

Close of day
Waterlilies close their petals as the daylight fades.

Welcome lights
As evening falls, harbour lights guide boats safely back to shore.

k

Evening

Hanging around
Bats rest between their nightly hunting trips by hanging upside down.

Night

Sleepy-head

You will spend about 25 years of your life fast asleep!

Big Moon

The Moon looks enormous, but it's only about as wide as Australia.

www.dk.com

Editor Jane Yorke
Senior Editor Mary Atkinson
Senior Art Editor Chris Scollen
Art Editor Mary Sandberg
DTP Designer Phil Keeble
Production Josie Alabaster
Jacket Design Joe Hoyle
Picture Research Jamie Robinson and Lee Thompson

Published in Great Britain by
Dorling Kindersley Limited,
9 Henrietta Street,
London WC2E 8PS

2 4 6 8 10 9 7 5 3 1

A CIP catalogue record for this book is available from the British Library.

ISBN: 0-7513-5848-7

Colour reproduction by Colourscan, Singapore
Printed and bound in Italy by L.E.G.O.

Contents

Day index

DAY

Explore the world in bright sunlight

Claire Llewellyn

DK

London • New York • Sydney • Moscow • Delhi

The morning is the start of a brand-new day.

Morning light
Each morning, the Sun climbs high in the sky. It shines on busy cities and quiet country fields.

Alarm call
Many birds and animals wake at dawn. Noisy cocks begin to crow.

Afternoon

Shadows get longer in the afternoon, as the Sun slowly sinks.

Good morning!

We wake up after a good night's sleep, ready for another day.

The Sun gives energy for life on Earth.

Sun-loving flowers
Plants need sunlight to grow.
A sunflower gently turns its head
so that it's always facing the Sun.

Leaves soak up sunlight.

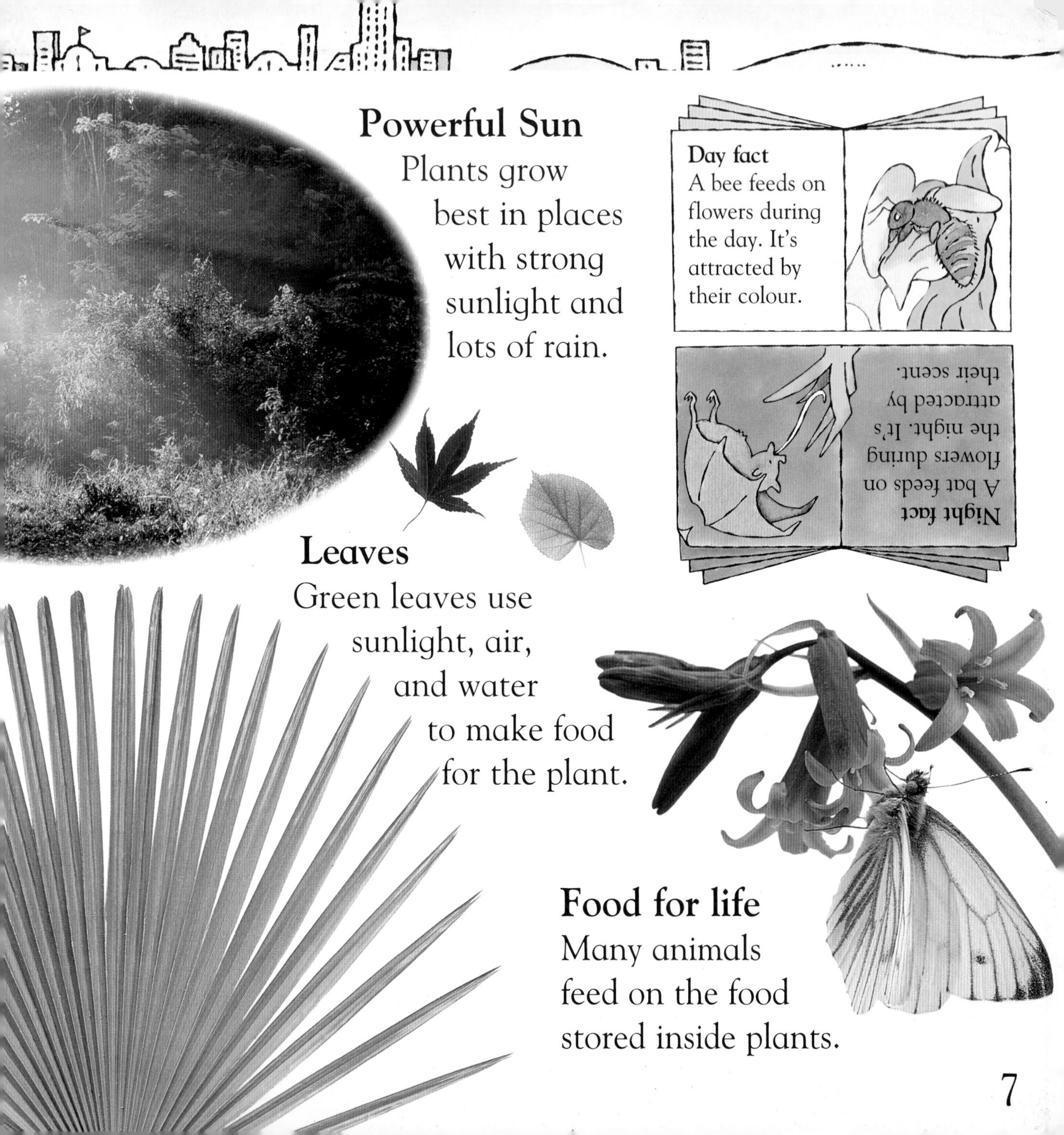

Powerful Sun

Plants grow best in places with strong sunlight and lots of rain.

Day fact
A bee feeds on flowers during the day. It's attracted by their colour.

Night fact
A bat feeds on flowers during the night. It's attracted by their scent.

Leaves

Green leaves use sunlight, air, and water to make food for the plant.

Food for life

Many animals feed on the food stored inside plants.

Daytime animals look around for danger or food.

The bird's long bill can reach ripe fruit.

Eye-spy

A toucan feeds in the daytime. It needs plenty of light to see in the rainforest.

Can't see me!

Animals are easily caught in daylight. This toad is using camouflage to hide.

Day fact
Dogs are active during the day. That's when they like to play.

Night fact
Cats are out and about at night. That's when they like to hunt.

Looking for danger

An impala has eyes on the side of its head. The whole herd keeps a lookout for hungry lions.

Sunbather

Lizards need sunshine to get them moving. They grow sluggish in the cold.

Busy people use the day to work, relax, and play.

A hairdresser

An office worker

A day's work

Most people go to work during the day. A job earns money to buy food, clothes, and a home.

Fitness and fun

Days aren't all work. Sport is fun and keeps your body fit too.

Sun and shade

Your shadow follows you on a sunny day but disappears at night.

Sunrise

At first light, plants start to make food for themselves.

Three meals a day

The day's mealtimes are a chance to relax, talk, and eat good food.

Off to school

On weekdays, children go to school, where they learn and play with friends.

Could an owl hunt in the day?

No. An owl feeds on mice and voles, which only come out at night.

Lift the flaps to compare life in the daytime with life at night.

Midday meal

In your lifetime, you will probably enjoy more than 25,000 lunches.

Dawn chorus

Birds sing sweetly at dawn, warning rival birds to stay away.

Could a honey-bee feed at night?

No. A honey-bee feeds on colourful flowers, which it can only see in daylight.

Lift the flaps to see the differences between night and day.